THE VOIDLESS UNIVERSE

A DISCUSSION OF SOME CURRENT COSMIC "MYSTERIES"

By Bob McClure V

Original text written solely by
Robert McClure V

Cover Photograph by
NASA Hubble Heritage Team

Mister Five Publications

The Voidless Universe
A Discussion of Some Current Cosmic
"Mysteries"

Written and published by Robert McClure V

Cover: photograph by NASA Hubble Heritage Team

First printing: 2008
Second printing: 2013
Third printing: 2015
Fourth printing: 2016
Fifth printing: 2018

Photo and Drawing Acknowledgements:

Cover: NASA Hubble Heritage Team
Page 0:   Illustration by WebCover
Page 3:   Figure 1 – Scienceworld.wolfram.com
Page 5:   M100 – HST Team
Page 12:   Figure 5 – Goddard Space Flight Center
Page 13:   Figure 6 – Hubble Space Telescope (HST) Team
Page 17:   M100 – HST Team
Page 18:   M63 – Adam Block / M101 – Hubble Heritage Team
Page 19:   M65 – Wendel / M109 – Adam Block
Page 20:   NGC 1365 – ESO / NGC 300 – ESO
Page 21:   M51 – HST Team / M91 – Joseph D. Schulman
Page 22:   M77 – Adam Block / M99 – AURA/NOAO/NSF
Page 23:   M88 – NOAO/NA/Sharp
Page 28:   Wikipedia – Lambda CDM Model
Page 30:   NGC 1316 – HST Team
            Hubble Extreme Deep Field – HST Team
Page 34:   Wikipedia – Cosmic Background Explorer
Page 35:   Wikipedia – Wilkinson Microwave Anisotropy Probe
            Wikipedia – Two Micron All-Sky Survey (2MASS)
Page 37:   NGC 4038 – HST Team
Page 43:   3C273 – Spitzer Space Telescope Team

Mathematical Acknowledgements:

Page 4:   Scienceworld.wolfram.com – Gravity
Page 6:   Scienceworld.wolfram.com – Chord
Page 13:   Wikipedia – Logarithmic Spiral
Page 14:   Wikipedia – Moment of Inertia
            Wikipedia – Angular Momentum
Page 32:   Wikipedia – Hubble's Law
            Wikipedia – Gravitational Potential
Page 36:   Wikipedia – Planck's Law
Page 39:   College Physics Handbook – Celerity
Page 41:   Wikipedia – Gauss' Law
Page 52:   Wikipedia – Occam's Razor

Experimental sciences, such as Chemistry and Physics frame their work within the scientific method: conduct their experiments, record the results, and then publish or otherwise circulate their discoveries, if any. Astronomy, however, is an <u>observational</u> science which must work on another plane. Astronomers are charged with observing the Universe in its natural state and then developing theories to explain what has been seen. Most observations easily fit well-understood concepts or theories. Then again, many others do not. Some of <u>those</u> cases are eventually resolved as being complex results of several simpler physical influences whose actions compounded themselves in a single situation.

Then there are some newly-discovered cases where Nature just doesn't seem to conform at all to previously accepted concepts. Observational and theoretical astronomers are moved to provide some new theory that hopefully will explain "the Problem". The current onrushing revolution in observational technology is compiling mountains of images and data about the Cosmos at all distances, levels of complexity, and degrees of interaction. A growing number of those observations are confounding rational attempts to explain them – at least according to the usual schools of thought.

Remembering that Nature is the master and that mathematics is only a tool – we'll entertain a new "school of thought" and see if rational review of its vision of the Cosmos can bring us closer to solving some of the more puzzling problems of Astronomy's last several decades.

SOME PROBLEMS

- Gravity doesn't seem to be strong enough to hold together spiral galaxies like the Milky Way. Why do they exist?
- Spiral arms rotate at speeds that don't change much with distance from their galactic centers. How is that possible?
- The "Winding Problem" – after dozens of rotations, why haven't the spiral arms simply wound themselves around their galactic cores?
- Why do galaxies in the Hubble "Deep Field" studies look so much like the ones nearby? Shouldn't they all be much younger looking?
- Why does the Universe contain more iron and heavier elements than it should? Why do many ancient quasars have "modern" levels of iron in them?
- Is the cosmic expansion of the "Big Bang" really accelerating? How? Why?
- Dark matter and dark energy – so hard to find. Do they really exist? Should we keep looking for them?
- Is the speed of light really the highest velocity possible? What is it – actually?
- What is it – actually, that gravity is really attracting?

A NEW VIEW OF THE UNIVERSE

Let's start by defining the term "voidless". It means more than "space is not empty". It means that the planets, stars, galaxies, molecular clouds, plasmas – everything in the universe – touches everything else in the cosmos by expressing its identity on the space around itself – out to an infinite distance. What in the "old school" is thought of as the empty space between separate celestial bodies traveling through it is now to be considered a region of greatly lower density – but still part of the same all-pervasive continuum. Celestial bodies themselves are actually volumes of highly concentrated mass-energy density. The sun of our solar system expresses its identity continuously – right through the cores of the Earth and other planets – and out to the rest of the Milky Way and the Universe, though at a strength which reduces according to the standard formulas. Extending this idea, "void-less" also means that forces and fields expressed in the continuum "superimpose" them-selves onto the reference frames through which we observe and measure them. This can be extrapolated upwards to gain a view of the entire Cosmos' effect on itself.

This concept provides new capabilities for understanding the force of gravity and helps solve the problems noted above. But first, we must go back to the most basic description of gravity and skip the shortcut usually taken – which has led to many of the observa-tional problems noted above.

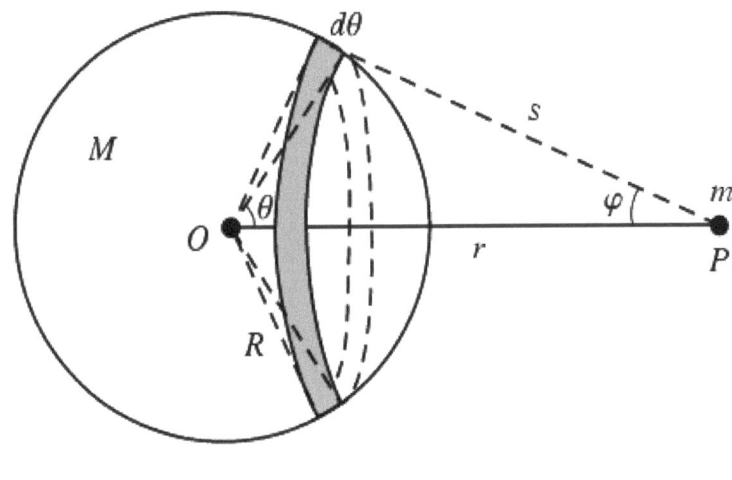

Fig 1

GRAVITY FROM A DISTRIBUTED SOURCE

As might be expected, deriving the strength of gravity arising from a "distributed source" rather than between "point sources" is going to mean talking about geometry – the geometry of spheres, the geometry of disks, and the geometry of spherical angles. The diagram above was used by Isaac Newton to derive his point-source formula for gravity.

He defined a massive sphere as a series of concentric spherical shells extending from the origin out to a radius R. Considering one shell, he integrated the differential mass elements defined by spher-ical angles (Θ and φ) across the shell's entire surface. The incre-ment of force created by the shell element dM felt by the test mass (little m) at a distance r is:

$$dF_r = -\frac{G*m*dM}{s^2}\cos\varphi \qquad [1]$$

Of particular interest for us in this equation is the "dM" element of mass. For spherical sources of gravity Newton continued his derivation, simplifying the equations and integrating the force through the entire radius of the massive sphere to achieve his famous formula:

$$F = -\frac{GMm}{r^2} \qquad [2]$$

He followed with a cautionary statement that the formula would be fully accurate only for those situations where the objects under study were spheres (or very nearly so) and also existed far enough from each other to act as "point sources". It is the historical failure of astronomers and physicists to remember this warning that has led to an inability to resolve the issues listed above. Evaluating the strength of gravity at places where spherical geometries **do not prevail**, such as the outer regions of a spiral galaxy, is a principal aim of this discussion, so we'll frequently need to use the more general differential formula shown in equation [1].

Even a casual glance at a spiral galaxy gives the sense of a great turning pinwheel. Angular momentum seems to keep its signature clouds of dust and gas (and the billions of young, massive stars they create) suspended above the core. Ever since spectroscopy first showed that spiral galaxies rotate, a big mystery has presented itself – so far without plausible explanation. Why are spiral galaxies' arms "frozen" in these rather fantastic shapes, instead of twisting ever more tightly around their central cores – which would then become elliptical galaxies? (The Milky Way has completed at least forty rotations during its lifetime, yet its principal arms still maintain their basic shape and structure.) Also mysteriously, spiral arms generally rotate at speeds that are constant with respect to distance from their galactic center.

Standard theories of gravitational attraction cannot accommodate this systemic motion, unless the entire galaxy were embedded in a massive halo of "dark" material – that has never been observed. After nearly fifty years of debate unhindered by substantiating observation, space-age telescopes finally provided a false clue. Low-temperature "deep" pictures of spirals **do** show sizable clouds of usually unseen matter surrounding the luminous parts of their galaxies. Is this "halo" material massive enough to constrain the galaxy's angular momentum?

"Deep" view of M100

The answer is *NO*. The usual view of Newtonian gravity requires these shrouds to comprise 90% of the galaxies' total material in order to constrain the luminous galactic spiral arms. The measured sizes of the non-luminous material and galactic "extinction" values (the ratio of drops in density) at the ends of the luminous regions indicate that only about 2% of an average galaxy's mass is hidden in such a dark shroud.

To start off solving this riddle, let's look into an interesting neighborhood: the region between a spiral galaxy's central core and its arms. The force felt from the core will behave according to Newton's familiar formula, even for our test object close to the "surface" of the enormous spherical mass. But what about that considerable mass of material (both luminous and non-luminous) hanging just overhead in a flattened disk? How much of an attraction is pulling from *that* direction – toward the outer reaches of the galaxy's arms?

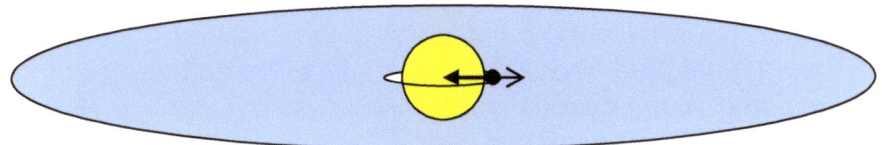

Fig 2

Densities in a spiral galaxy change sharply at discrete boundaries. Density of spiral arm material is usually about 10% as great
as the central core. Extinction values for many spirals indicate that
the halos of non-luminous material surrounding the galaxies have
about 0.1% – 1% of the density of the spiral arms. So we have two
distinct disk geometries to consider, but both computations follow
the logic of the diagram just below.

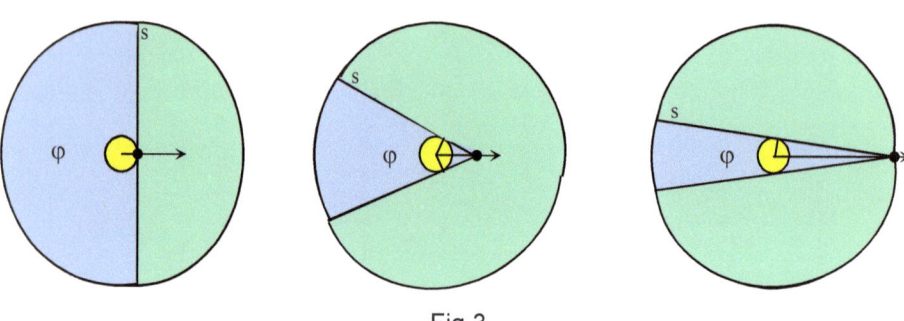

Fig 3

Lines tangent to the surface of the galaxy's central core are
drawn from the test object's location. Mass in the disk that is further
away from the center than these lines will attract the test object toward itself with a resultant vector sum aimed directly away from the
galactic core, in the plane of the disk. We can measure the strength
of this force by integrating Newton's differential formula:

$$F = -Gm \int \frac{\cos \varphi * dM}{s^2} \qquad [3]$$

dM is given by the disk's density times its thickness times the
area considered.

$$Area_{disk} = \frac{4sd}{3} + \frac{d^3}{s} - \frac{\left[(2r-s)^2 * \left(\frac{\pi}{2} - \varphi\right)\right]}{2} \qquad [4]$$

Here, **d** is the diameter of the disk's segment and **r** is the distance from our measuring point to the galactic center.

The other neighborhoods of interest are at the boundaries of the
luminous and non-luminous spiral arms. Information relating to
these areas is included in table 1:

6

	Mass (kg)	Radius (m)	h (m)	den (kg/m³)	Surface Area (m²)	Edge Mass (kg)
Core:	1.85×10^{40}	5.60×10^{19}	5.6×10^{19}	1.29×10^{-20}	3.94×10^{40}	5.08×10^{20}
Lum. Disk:	7.74×10^{39}	2.58×10^{20}	1.91×10^{19}	1.29×10^{-21}	2.09×10^{41}	3.99×10^{19}
NonLum Disk:	4.00×10^{38}	3.49×10^{20}	5.72×10^{19}	2.38×10^{-23}	3.83×10^{41}	2.99×10^{18}

Table 1

The boundary volume and mass figures shown on the second line in the table above are constrained to the thickness of the galactic disk and describe the mass and forces felt in a one-meter-thick "peel" of the galaxy's edges at our three ranges of interest. Non-luminous disk matter is constrained to a disk three times the thickness of the luminous disk material and is presumed to occupy space above and below the luminous disk – as well as extending beyond it in range from the galactic center. This is in consonance with observations and theoretical studies.

How does the central core's gravity compare to this new "outward" gravity and the "centrifugal" effect of the galaxy's rotation? Our "average galaxy" rotates with a speed of 225.58 kilometers per second. The "centripetal inertia" acceleration thus produced and the "outward" gravitational acceleration (formulas [8] & [9] below) are compared at these three distance ranges to the inward accelerations due to gravity.

	Gravity (Outward)	Spin (Outward)	Total (Outward)	Gravity (Inward)
Core:	3.0450×10^{-12}	9.0888×10^{-10}	9.1193×10^{-10}	3.9595×10^{-10}
Lum. Disk:	1.2349×10^{-12}	1.9738×10^{-10}	1.9861×10^{-10}	2.1139×10^{-11}
NonLum Disk:	1.0201×10^{-12}	1.4571×10^{-10}	1.4673×10^{-10}	1.1569×10^{-11}

Table 2

We can see at all three distance ranges, the inward attractions on disk material are not strong enough to hold it in place. Does this mean that matter is always being lost by the galaxy as it pinwheels through space? While that *is* a possibility, the stable population of spiral galaxies (35% of regular matter in the Universe) indicates otherwise. Let's try again, this time by determining the strength of galactic gravity in the spiral arms as a function of radial distance from the core's center.

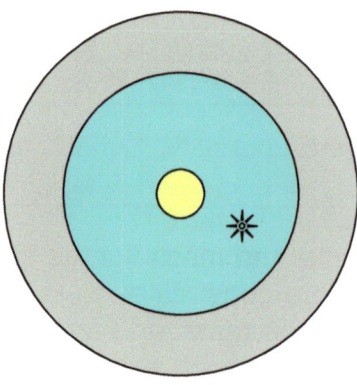

Fig 4

We want to show that galactic arms "freeze" their shape by at-tracting nearby matter within themselves at least as strongly (in the opposing direction) as they are attracted by the galactic core.

In the diagram just above, we're examining a point in the lumi-nous portion of the galactic disk. We'll consider that the core (in-nermost circle) is a sphere ten times as dense as the disk. The non-luminous matter (outermost band) is a disk three times as thick but only one percent as dense as the spiral arm material (middle band) – so it also pervades the regions above and below the lumi-nous disk. For simplicity's sake, we'll assume the galaxy's density to be constant within each of these three geometric portions. To avoid another dimensional complication, let's assume our test mass to be on the midline of the disk's material. In actual fact, spiral arm material "twists" above and below the midline as it rotates around the galactic core, but since we are concerned with the overall spiral shape over time, this approximation is justified. Also, our "test ob-ject" won't have mass, so we'll discuss the acceleration it will feel, rather than the force on an object with mass.

To see what accelerations act on our test object, let's start at the object's location and move through the galaxy's mass distribu-tion, adding terms as we go. At the object itself, we have the gal-axy's rotation to consider, so:

$$a_o = \frac{v^2}{r}$$ [5]

where the sample average v = 225.58 $^{km}/_{sec}$. This equates to:

$$g_o = \frac{5.0884 * 10^{10}}{r} \; \text{m/sec}^2. \tag{6}$$

The next gravitational acceleration felt will arise from the luminous disk material. Acceleration vectors appearing in this galactic section must be derived from Newton's integral formula:

$$g = -G \int \frac{\cos\varphi * dM}{s^2} \tag{7}$$

where s = R*(1+cosφ).

The **outward** gravitation contribution from <u>luminous</u> disk material, as a function of distance from galactic center, is:

$$g_1 = \left(2.3253 * 10^{-10}\right) * \frac{4sd}{3} + \frac{d^3}{s} - \frac{\left((2r-s)^2 * \left(\frac{\pi}{2} - \varphi\right)\right)}{2} \tag{8}$$

Similarly, outward acceleration due to the <u>non-luminous</u> disk is determined by:

$$g_2 = \left(8.3013 * 10^{-10}\right) * \frac{4sd}{3} + \frac{d^3}{s} - \frac{\left((2r-s)^2 * \left(\frac{\pi}{2} - \varphi\right)\right)}{2} \tag{9}$$

These accelerations all point in the direction directly outward from the galactic core. They are opposed by the gravity of the core and also that of the disk matter closer to, and on the far side of, the galactic center. For the core, this numerically evaluates as:

$$g_3 = -\frac{1.2408 * 10^{30}}{r^2} \; \text{m/sec}^2 \tag{10}$$

For disk material pulling our test object toward the galactic core, a precise solution would be to subtract the areas for the g_1 and g_2 values from the area of the entire galaxy. Multiplying by the same corresponding numeric values for luminous and non-luminous distributions would give their contributions to the inward acceleration. Though this element is significantly smaller than the inward attraction of the galactic core (as shown in graph 2 below), it contributes to the "ring maintenance" effect of the spiral arm material on itself.

Salient measurements were taken for twelve galaxies of common astronomical interest: NGC 4321; NGC 5457; NGC 5505; NGC 3992; NGC 3623; NGC 300; NGC 1365; NGC 4548; NGC 5194; NGC 1068; NGC 4254; and NGC 4501. From these twelve galaxies average values were calculated for size, mass, rotation speed, disk thickness, and core radius. Stellar extinction values and non-luminous matter density estimates were also averaged.

Graph 1 compares the inward and outward accelerations experienced by galactic material as a function of radial distance from the galaxy's center. It shows that the outward accelerations holding spiral arm material away from the galactic core are greater at every range of distance than the accelerations toward it.

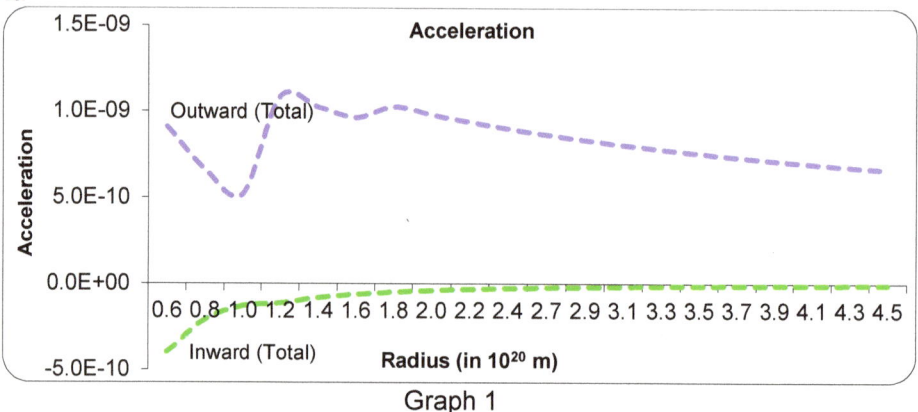

Graph 1

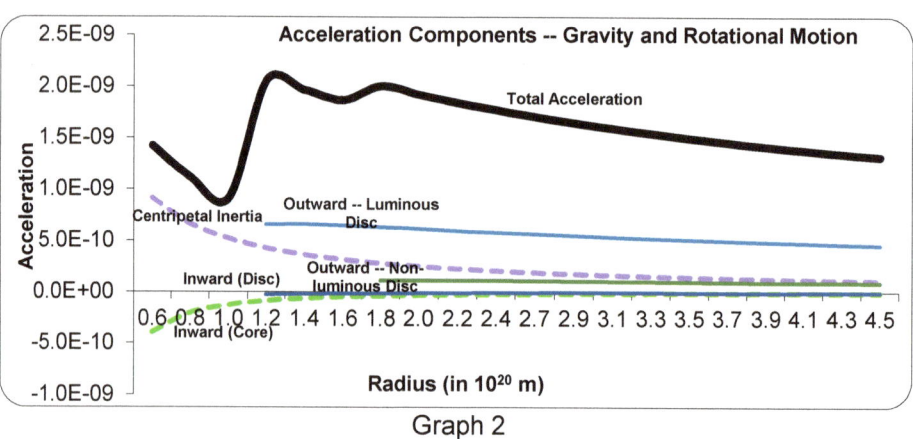

Graph 2

Graph 2 breaks down the accelerations into five components: two inward which strive to keep the galaxy together – and three outward components which work toward dispersing the galaxy across the Cosmos. The black line shows the sum of all five.

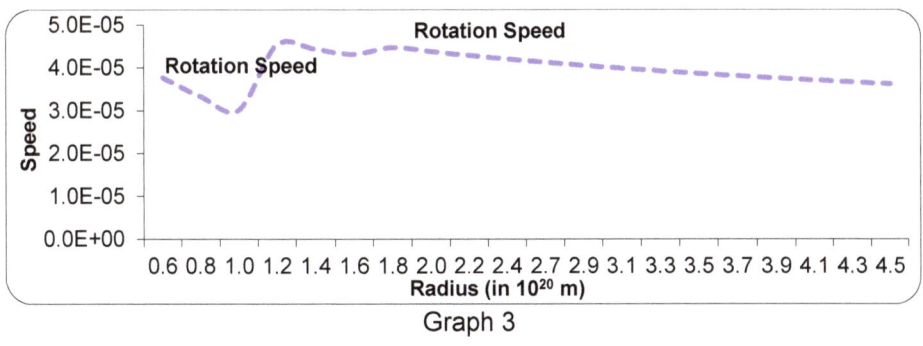

Graph 3

Graph 3 shows the square root of both the accelerations displayed in graph 1. It tracks very closely with observed galactic rotation curves. This demonstrates that outwardly-directed gravitational attraction of the disc material on itself eliminates both the "Winding Problem" and the "Constant Velocity Problem".

> **ASSERTION 1**: The material in galactic spiral arms is more significantly influenced by the matter near it in the spiral disk than by the galactic core. A galaxy's spiral arms (and associated bars) are actually a separate gravitational system.

The galactic core is the heart of the galaxy, to be sure – but its gravitational influence on spiral arm material is strongly overcome by the spiral arm material's influence on itself. We should not think of galactic arm material as "rotating" around the galaxy's axis – but rather consider it to be "revolving" around the galaxy's core – like the rings of Saturn.

But since the "outward" accelerations experienced in the spiral arms are so much stronger than gravitational attraction toward the core, the first question raised above (Why do spiral galaxies still exist?) has just been made harder to answer – or so it would seem. Indeed, it is interesting to note that our "average" galaxy would not stay together even down to its core, given the rate of rotation noted – when the gravity between its components are treated in the usual point-to-point Central Force method of computation. We need to work more deeply on the general nature of galactic geometry to determine if there isn't something we've missed in our understanding of galactic gravity.

GALACTIC GEOMETRY – GENERALIZED

The apparent complexity of the Universe arises not from its rules (gravity, momentum, inertia, thermodynamics, etc.), which are fairly simple, but rather from the application of those rules to a wide variety of circumstances. As the Universe first coalesced into immensely massive aggregates of material we now know as galaxies, random probability provided the wide range of differences we observe in galactic masses – and also galactic momenta, both linear and angular. Can these simple elements of Physics account for the menagerie of galactic shapes observed in space? Do they "evolve" from one shape to another? If so, how? And in which direction?

Elliptical galaxies dominate those regions of the Cosmos where the density of galaxies is somewhat higher than the overall norm. Spiral galaxies (and irregulars) are more prevalent in the "less dense" neighborhoods. Computer modeling of what happens to spiral and irregular galaxies when two or more galactic objects merge firmly indicates that galactic evolution heads toward the elliptical geometry. (From right to left in figure 5.) This is also substantiated by the fact that elliptical galaxies are generally much larger and more massive than spiral galaxies. This is to be expected as the history of the Universe unfolds and galaxies cannibalize each other under the constantly attractive force of gravity. A higher population of elliptical galaxies in regions of higher density is a logical outcome of the merging process.

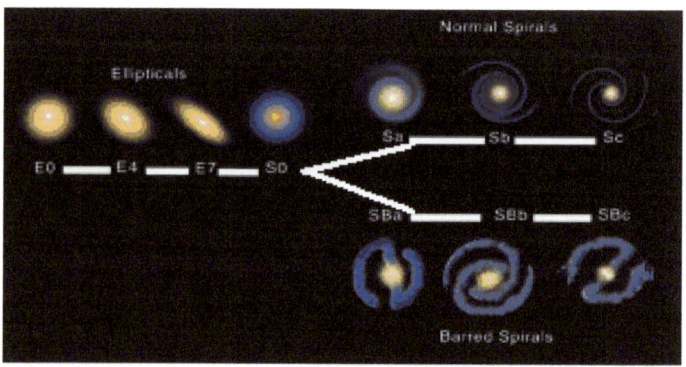

Fig 5

So what can we say about the three shapes of galaxies –spirals , barred spirals and ellipticals? Is there one generalized equation that can describe all these shapes?

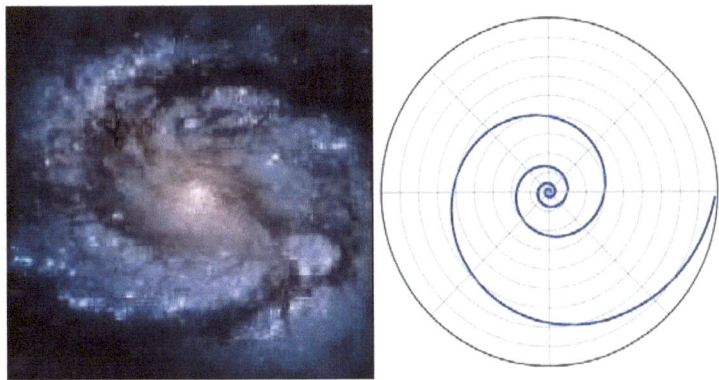

Fig 6

When not disturbed by collisions, the natural form taken by all galaxies is that of the "logarithmic spiral". (See Figure 6.) The polar coordinate expression for this shape is:

$$r = ae^{b\theta}$$ [11]

Applying this mathematical curve to galactic circumstances is a matter of determining appropriate values for the two terms "a" and "b". When "b" is very small, the spiral approaches a circular curve of radius "a". When "b" grows towards infinity, the spiral stiffens out toward the shape of a straight line. Intermediate values of "b" establish the graceful curves we associate with spirals –determined by what is known as the "pitch", or "growth" factor of the spiral.

The initial variables for all galaxies as they condense from the primordial clouds of their creation are their mass and their momentum – both linear and angular. A galaxy's mass and density will dictate its size (the "a" variable). Its linear momentum will be pertinent to the galaxy's existence in a group, cluster, or supercluster of galaxies. Its angular momentum is the physical characteristic that determines its shape.

What we've determined just now is that Hubble's "Tuning Fork" diagram of galactic evolution should be a straight line, with a galaxy's angular momentum determining its morphology.

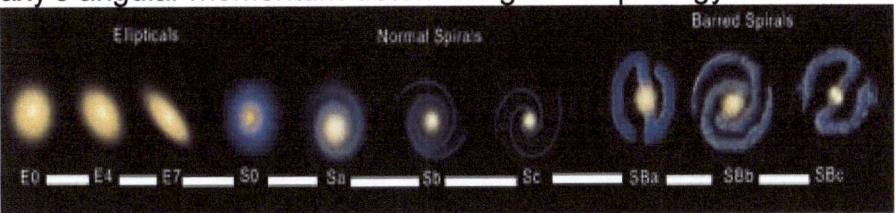

Fig 7

Though their constituent stars may swarm furiously around within them, elliptical galaxies exhibit little overall rotation or angular momentum. So the values of their "b" factors are small – essentially zero in the "E0" galactic category shown in figure 7.

Without doubt, the values of angular momentum possessed by galaxies as they first formed were spread between zero and some maximum above which the constituent galactic material would spin off into other neighborhoods of the Universe. The simplistic view of Newtonian gravity as shown in formula [2] says that nearly all spiral galaxies are already in such a condition – which prompted early theories about the existence of still-never observed "dark matter" as a possible explanation. But since about 35% of the Universe's visible matter exists in spiral galaxies, we know there must be other factors that create stability for this form.

What we want to establish now are the terms for the curve in equation [11] that will provide physical meaning at the galactic scale. As noted above, the factor "a" can be defined as the radial distance from the galactic center of the component under consideration. The "growth" factor (the "b" term) which provides the curve of the spiral arms is determined by its angular momentum:

$$L = I * \omega \qquad\qquad [12]$$

Where ω is the angular velocity of the galaxy and I is its moment of inertia, which operates as mass in rotational circumstances. In situations such as this where the radius of revolution or rotation is much greater than the translational speed, angular velocity can be approximated by:

$$L = M * v * r \qquad\qquad [13]$$

Spiral galaxies are generally comprised of four major components: a spherical core, a central bar (now seen in at least 60% of spiral galaxies), the spiral arms, and a spherical halo of stars and other material. We start by determining the angular momentum of the galaxy's rotating components:

$$L_{Core} = \frac{2}{5} * M_C * v * R_C \qquad\qquad [14]$$

$$L_{Bar} = \frac{1}{12} * M_B * v * R_B \qquad\qquad [15]$$

$$L_{Disc} = \frac{1}{2} * M_D * v * R_D \qquad\qquad [16]$$

14

A note to consider: Comparison of formulas [15] & [16] leads to an interesting observation regarding the logarithmic spiral in equation [11]. High values of "b" are necessary for the spiral to "flatten" toward the straight line shown in galactic bars. But only one sixth as much rotational momentum is required to form a bar structure as that required for a disc. Conservation rules dictate that the material's spin will increase as matter gathers closer together and "R" decreases. [This also raises the curve's "b" factor.] These two influences work together to establish that spiral galaxies will have bar structures within them whenever their initial angular momentum is a notable factor in their formation. This is why so many galaxies, including our own Milky Way, are found to contain bars within their spiral structure.

A galaxy's pitch angle is governed by its angular momentum. The growth factor "b" is the galactic component's angular momentum divided by its mass. Pitch angles were identified for four galaxies from our sample of twelve and subjected to formula [11] as shown above. Formulas [14], [15], and [16] were used to derive values for angular momentum and mass for each of the rotating components. The growth factors so derived were consistent to within 2% across all three galactic components for all four galaxies. This demonstrates that galaxies (including ellipticals and barred spirals) do indeed follow the logarithmic spiral shape determined by formula [11].

Up to this point, we've discussed galactic gravity in terms of acceleration and force (which is mass times acceleration). To proceed in our discussions about galactic shapes (and other astrophysical issues) we shift to talking about energy – both kinetic and potential. We've discussed galactic kinetic energy in terms of the galaxies' motion through space and their angular momentum.

Explaining how spinning galaxies maintain their existence is a matter of determining their gravitational potential energies. In general terms, energy is defined as the integral of force:

$$U = \int F dr \qquad [17]$$

where Newton's original definition of gravity provides:

$$F = -Gm \int \frac{\cos \varphi * dM}{s^2} \qquad [18]$$

In this case we'll use the usual formula for gravity, but still keep in mind Newton's cautionary remark about "spherical geometry" and "point sources" – and make adjustments if necessary. So we start with:

$$F = -\frac{GMm}{r^2},$$ [19]

Which leads to the integral definition of energy:

$$U = -GMm\int \frac{dr}{r^2}$$ [20]

And this finally evaluates as:

$$U = -\frac{GMm}{r}$$ [21]

Graphing this function with respect to "r", the distance from the center of a massive body, draws a picture of what physicists usually call the "Gravitational Potential Energy Well", as shown in figure 8. The energy a test object would need to escape the massive body is measured by how far its position on the curve sits below the horizontal line at the top. As long as the object's position is below the line, it feels some gravitational "force" from the massive body. In his General Relativity Theory, Einstein correctly predicted that this "field effect" produced on space by the presence of a massive body would be able to bend a beam of light. The quantitative effect of such a field is discovered by determining how much energy would be required by a test object to escape from the massive body.

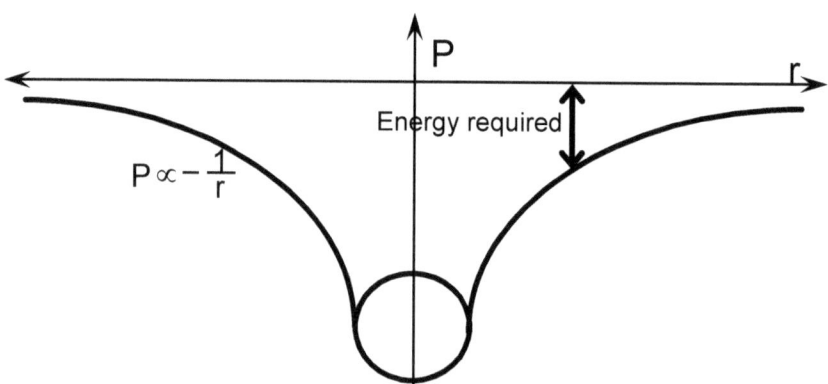

Gravitational potential, expressed as function of a test object's distance from the massive body.

Fig 8

Gravitational potentials are used to discuss strengths of fields because they focus on the "work" capacity of the field. We want to determine whether the gravity exerted by a spiral galaxy can work hard enough to constrain material at its outermost boundaries.

The error compounded through the 20th and 21st centuries is to consider gravity working within a "test mass responding to a central force" context, which fosters only point-to-point computations. As Newton predicted, this works very well for spherical and small bodies moving at significant distances from each other. But the computation breaks down when galactic geometries are involved. In 1933, this led the astronomer Fritz Zwicky to propose the existence of huge clouds of never-observed "dark matter" shrouding galaxies – sometimes comprising 80% or more of a galaxy's required mass.

As the "deep" view of M100 on page 5 shows, there <u>are</u> clouds of non-luminous material surrounding spiral galaxies. But their extent and lower densities (numerically defined below as "extinction" values) are shown to make up less than 3% of our average galaxy's mass.

The figures below refer to the gravitational potentials [in Joules] felt by a one-meter-thick "peel" of galaxies at the outside edges of three significant boundaries.

M100 (NGC 4321)

	Core	Luminous Arms	Non-Lum Arms
Radius [m]:	1.2344×10^{19}	2.6838×10^{20}	4.0257×10^{20}
Mass [kg]:	1.0159×10^{38}	8.0868×10^{39}	9.1648×10^{38}
Rotation Speed [m/sec]:	2.836×10^5	Disk Thickness [m]:	2.777×10^{19}
Extinction Value:	0.01677		
Closure Rqmt [J]:	5.082×10^{28}	1.0326×10^{29}	2.5753×10^{27}
Gravitation Pot. [J]:	1.3561×10^{28}	1.2297×10^{29}	6.5716×10^{27}

M63 (NGC 5505)

	Core	Luminous Arms	Non-Lum Arms
Radius [m]:	7.4681×10^{19}	5.7606×10^{20}	8.6409×10^{20}
Mass [kg]:	2.2497×10^{40}	4.9352×10^{40}	4.2585×10^{39}
Rotation Speed [$^m/_{sec}$]:	1.721×10^{5}	Disk Thickness [m]:	3.7341×10^{19}
Extinction Value:	0.012663		
Closure Rqmt [J]:	1.4624×10^{29}	1.0798×10^{29}	2.0338×10^{27}
Gravitation Pot. [J]:	4.5422×10^{30}	1.4506×10^{30}	5.8376×10^{28}

M101 (NGC 5457)

	Core	Luminous Arms	Non-Lum Arms
Radius [m]:	4.9685×10^{19}	2.1834×10^{20}	2.6201×10^{20}
Mass [kg]:	6.6245×10^{39}	1.0737×10^{39}	5.6729×10^{37}
Rotation Speed [$^m/_{sec}$]:	2.742×10^{5}	Disk Thickness [m]:	5.8634×10^{18}
Extinction Value:	0.01203		
Closure Rqmt [J]:	3.9128×10^{28}	1.6652×10^{28}	2.3944×10^{26}
Gravitation Pot. [J]:	2.1002×10^{29}	2.4406×10^{28}	8.8727×10^{26}

M65 (NGC 3623)

	Core	Luminous Arms	Non-Lum Arms
Radius [m]:	1.2437×10^{20}	2.026×10^{20}	2.0361×10^{20}
Mass [kg]:	1.0389×10^{41}	1.9826×10^{39}	3.9488×10^{37}
Rotation Speed [m/sec]:	2.331×10^{5}	Disk Thickness [m]:	1.9133×10^{19}
Extinction Value:	0.006534		
Closure Rqmt [J]:	2.2639×10^{29}	3.6496×10^{28}	2.3963×10^{26}
Gravitation Pot. [J]:	1.0748×10^{31}	1.0953×10^{30}	2.1479×10^{28}

M109 (NGC 3992)

	Core	Luminous Arms	Non-Lum Arms
Radius [m]:	4.9685×10^{19}	2.22×10^{20}	3.3301×10^{20}
Mass [kg]:	6.6245×10^{39}	1.4632×10^{39}	3.8142×10^{38}
Rotation Speed [m/sec]:	2.951×10^{5}	Disk Thickness [m]:	7.715×10^{18}
Extinction Value:	0.03752		
Closure Rqmt [J]:	5.9632×10^{28}	2.5794×10^{28}	1.4392×10^{27}
Gravitation Pot. [J]:	2.7634×10^{29}	3.3738×10^{28}	3.9766×10^{27}

NGC 1365

	Core	Luminous Arms	Non-Lum Arms
Radius [m]:	7.4681×10^{19}	3.4002×10^{20}	3.7402×10^{20}
Mass [kg]:	2.2497×10^{40}	1.0317×10^{40}	1.0942×10^{39}
Rotation Speed [m/sec]:	1.652×10^{5}	Disk Thickness [m]:	2.3145×10^{19}
Extinction Value:	0.028964		
Closure Rqmt [J]:	1.2022×10^{29}	5.298×10^{28}	1.6846×10^{27}
Gravitation Pot. [J]:	2.8154×10^{30}	4.1065×10^{29}	3.6872×10^{28}

NGC 300

	Core	Luminous Arms	Non-Lum Arms
Radius [m]:	1.2344×10^{19}	1.3983×10^{20}	2.0974×10^{20}
Mass [kg]:	1.0159×10^{38}	1.2126×10^{39}	1.4413×10^{38}
Rotation Speed [m/sec]:	8.97×10^{4}	Disk Thickness [m]:	1.543×10^{19}
Extinction Value:	0.017533		
Closure Rqmt [J]:	2.8244×10^{27}	3.032×10^{27}	7.905×10^{25}
Gravitation Pot. [J]:	8.4758×10^{27}	1.0964×10^{28}	6.3995×10^{26}

M51 (NGC 5194)

	Core	Luminous Arms	Non-Lum Arms
Radius [m]:	8.4556×10^{19}	1.7969×10^{20}	2.6055×10^{20}
Mass [kg]:	3.2654×10^{40}	2.3569×10^{39}	1.9589×10^{38}
Rotation Speed [$^m/_{sec}$]:	1.301×10^{5}	Disk Thickness [m]:	2.3145×10^{19}
Extinction Value:	0.011467		
Closure Rqmt [J]:	5.849×10^{28}	1.2229×10^{28}	2.0171×10^{26}
Gravitation Pot. [J]:	4.0864×10^{30}	4.3814×10^{29}	1.5156×10^{28}

M91 (NGC 4548)

	Core	Luminous Arms	Non-Lum Arms
Radius [m]:	5.74×10^{19}	1.2834×10^{20}	1.4759×10^{20}
Mass [kg]:	1.0214×10^{40}	8.2362×10^{38}	2.2407×10^{37}
Rotation Speed [$^m/_{sec}$]:	1.779×10^{5}	Disk Thickness [m]:	1.543×10^{19}
Extinction Value:	0.006463		
Closure Rqmt [J]:	4.9915×10^{28}	1.0967×10^{28}	8.1264×10^{25}
Gravitation Pot. [J]:	8.522×10^{29}	9.2091×10^{28}	1.7892×10^{27}

M77 (NGC 1068)

	Core	Luminous Arms	Non-Lum Arms
Radius [m]:	6.2337×10^{19}	2.2332×10^{20}	3.9081×10^{20}
Mass [kg]:	1.3084×10^{40}	2.8744×10^{39}	1.3932×10^{39}
Rotation Speed [$^m/_{sec}$]:	2.825×10^{5}	Disk Thickness [m]:	1.543×10^{19}
Extinction Value:	0.049913		
Closure Rqmt [J]:	1.3645×10^{29}	4.7551×10^{28}	4.1045×10^{27}
Gravitation Pot. [J]:	1.0916×10^{30}	1.3314×10^{29}	2.1677×10^{28}

M99 (NGC 4254)

	Core	Luminous Arms	Non-Lum Arms
Radius [m]:	3.9809×10^{19}	2.834×10^{20}	3.1174×10^{20}
Mass [kg]:	3.4076×10^{39}	7.3816×10^{39}	2.9439×10^{38}
Rotation Speed [$^m/_{sec}$]:	2.982×10^{5}	Disk Thickness [m]:	2.3145×10^{19}
Extinction Value:	0.010949		
Closure Rqmt [J]:	1.4708×10^{29}	1.0034×10^{29}	1.206×10^{27}
Gravitation Pot. [J]:	4.2644×10^{29}	1.3502×10^{29}	4.5559×10^{27}

M88 (NGC 4501)

	Core	Luminous Arms	Non-Lum Arms
Radius [m]:	2.9934×10^{19}	3.1162×10^{20}	4.6743×10^{20}
Mass [kg]:	1.4487×10^{39}	6.0137×10^{39}	4.6622×10^{38}
Rotation Speed [m/sec]:	2.722×10^{5}	Disk Thickness [m]:	1.543×10^{19}
Extinction Value:	0.011426		
Closure Rqmt [J]:	6.1826×10^{28}	6.1167×10^{28}	1.0394×10^{27}
Gravitation Pot. [J]:	1.2087×10^{29}	62259×10^{28}	2.2675×10^{27}

Average values compiled across the twelve sampled spiral galaxies are:

Averaged Galaxy	Core	Luminous Arms	Non-Lum Arms
Radius [m]:	5.5985×10^{19}	2.578×10^{20}	3.4921×10^{20}
Mass [kg]:	1.8596×10^{40}	7.7448×10^{39}	7.7193×10^{38}
Rotation Speed [m/sec]:	2.2558×10^{5}	Disk Thickness [m]:	1.9082×10^{19}
Extinction Value:	0.01852		
Closure Rqmt [J]:	9.6853×10^{28}	4.315×10^{28}	1.0755×10^{27}
Gravitation Pot. [J]:	1.9186×10^{30}	2.7177×10^{29}	1.5541×10^{28}

In each case and the averaged example, the Newtonian potential is sufficient to hold a spiral galaxy together at all three points of interest – when the energy levels are integrated across the entire range of their effectiveness. At the edge of the core of our averaged galaxy, the gravitational potential is 19.8 times greater than needed to constrain its material. At the edge of the luminous arms, gravity is 6.3 times greater than necessary. Out at the dark edge of the non-luminous (but still normal) matter, the galaxy's gravity is 14.5 times greater than needed to retain the galaxy's material.

> **ASSERTION 2:** Huge quantities of mysterious "dark matter" are **not** required for spiral galaxies to retain their material. Their massive cores, extensive luminous spiral arms and thin envelopes of <u>normal</u> non-luminous matter are sufficient to constrain all their material.

TO THE FARTHEST HORIZON

Looking up in the nighttime sky, we see thousands of stars spangled across the darkness. Astronomers peering through their telescopes examine the depths of space and find billions of stars suspended in the fathomless deep. The Universe we observe seems to be open, unbounded and infinite – when we just look at it. But is that really so?

The most widely accepted theory of Cosmology (once derisively called the "Big Bang" by one of its detractors), would establish limits to these apparent conditions. In the deep reaches of space, we find that the further out a particular object is from us, the faster it is flying away, as measured by the "red shifting" of the light we see. By conceptually retracing the paths and velocities of many distant celestial objects, astronomers saw them as merging together at a "beginning" of space and time somewhere between fifteen to twenty billion years ago. This placed a definite initial limit of TIME on the beginning of the Universe.

Another limitation generally placed on the Universe is SPEED, namely the velocity with which electromagnetic pulses travel (the speed of light). Applied to Cosmology, these two constraints establish another apparent limit on our Universe – DISTANCE. If the Universe has existed for only so long and objects within it can travel only so fast, then it should be fairly easy to derive an approximate maximum SIZE. If the "Big Bang" Universe initially inflated from one small kernel and light's speed is really the limit – it would be reasonable to construe the Universe's maximum size to be very close to the same number of lightyears as its age. But how old is that exactly – and is the relationship really linear?

A debate in Astronomy continued for years about the numeric value of something called the "Hubble Constant". It describes the rate at which almost all galaxies appear to be speeding away from our viewpoint here in the Milky Way. Accurately determining this number was so important that it was denoted as a "key project" for the Hubble Space Telescope (HST), when the big instrument was first launched into orbit. The first collation of observations for this purpose provided a figure that was embarrassingly high for astronomers to accept. An H_o value of 80 $^{km}/_{sec}$ per megaparsec meant the Universe could be no more than 8 to 10 billion years old. Many stars in the Milky Way's own globular clusters are known to be as much as 14 billion years old – based on extremely reliable stellar lifespan studies. So the Key Project was redone, with even more care than the first time. After two years more work, the results from the second run came to a value of 70 $^{km}/_{sec}$ per mpc (plus or minus 10%), which gave a "more comfortable" age of at least 12 billion years. Further refinements have recently narrowed down the estimate to around 13.75 billion years and the argument would appear to have been settled.

But recent theoretical studies regarding the Universe as a whole and what is termed the "Observable Universe" purport that the relationship between its age and its size is not linear, due to the expansion of space itself during the Universe's lifetime. Rather than a figure a bit smaller than 14 billion light years, these studies suggest that the Universe has a radius of at least 46 billion light years – and might actually be infinite after all.

The idea that space itself might expand three times faster than any material or energy it contains is, in itself, not necessarily a logical contradiction – but this "New Standard Model" of Cosmology leans on several illogical ideas for support.

First: the Universe is supposed to have come from nothingness to a golf-ball size in absolutely no time whatever.

Second: it has been inflating at a speed at least three times the speed of light ever since, primarily because its "yardstick" has been lengthening. [The "metric tensor" of General Relativity.]

Third: the expansion really has no center – the Universe is supposedly inflating from every point within itself.

The not-quite-exactly-equal expansion rate within the Universe purportedly gave rise to the distribution of galactic super-clusters, while the instantaneous nothing-to-golf-ball inflation supports the observed quantity ratio between hydrogen and helium in the Universe. Whether the Universe's history actually followed this storyline is a matter of conjecture – supported by some of the mathematics currently in use – but it is not really provable.

An even more fanciful construct – String Theory – is often touted as a way to explain the observed distribution of galaxies and the ratio of their primary material elements. It requires the existence of nine higher ordered, thus non-observable, dimensions that curl up within themselves more tightly than protons in an atomic nucleus. This is an obvious mathematical fantasy that possesses no real relation to actual fact. Going back to one of the original premises of this work – that Astronomy and Physics are be observational sciences wherein mathematics is one tool they may use – String Theory should be eliminated from any further serious consideration by astronomers and physicists.

Another illogical supposition of Astronomy and Physics perpetuated by the mathematics they use is a proposed imbalance between matter and anti-matter in the Universe. This "problem" is an illusion. It supposes that as matter was created from nothing at the beginning, there must have been an equal amount of anti-matter created that somehow just disappeared. One proposal suggests that vast amounts of matter / anti-matter were created together and that most of the material recombined into annihilation. Everything we have left in the Universe somehow just missed the big spasm of recombination. This is another example of how thought processes driven by mathematical theories rather than observations can lead to nonsense.

So what is really going on out in the Universe as a whole? What can our observations about galactic gravity tells us about it?

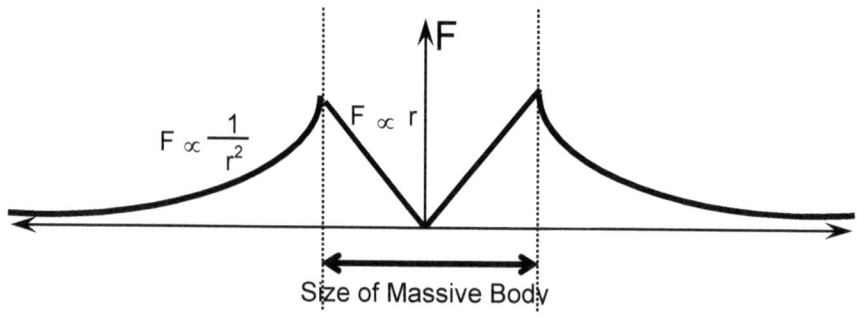

Strength of a massive body's gravity, expressed as a function of a test object's distance from its center.

Fig 10

To determine the true state of affairs, let's start by discussing a few more "basics" about gravity. Figure 10 is a graph of the strength of gravity expressed as a function of the distance from the center of a massive body. From the surface of the object outward, we see that the force obeys Newton's formula just fine. (Those curved slopes on the outside edges.) But that part of the graph inside the surface doesn't agree at all. What's going on – and what's the significance of the graph's value at the very center of the massive body? Is the gravity there really zero? If so, why?

It **is** zero because the entire mass of the body surrounds that point symmetrically, attracting a "test" object equally in all directions. The force cancels itself out exactly. Here's how:

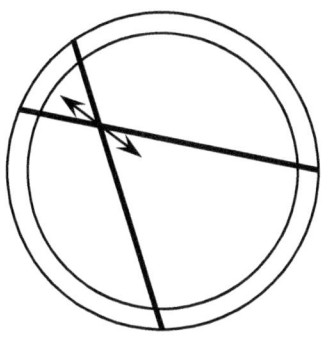

Fig 11

An interesting "thought experiment" is sometimes posed in Physics classes. It goes like this: If the Earth were not a solid body but hollow like an enormous basketball, with all its mass concentrated in a shell, its "external" gravity would be the same because of the "central force" effect demonstrated in Newton's derivation of his formula. But inside the shell there would be no gravity at all, anywhere – even for someone close enough to touch its wall. The figure just above shows why this happens. The formula for gravity indicates that the force <u>decreases</u> with the square of the distance from the source. Inside a spherical shell, the amount of source material generating the force (big "M") <u>increases</u> by the square of the distance and pulls from exactly the opposite direction. Thus, the combined effect of the whole shell results in a complete cancellation, no matter where within the sphere the test mass (little "m") is located.

Now consider our original massive body to be a series of concentric shells enwrapping each other. As our "test" object climbs up from the center, it "feels" a pull only from the mass below it. This volume grows in proportion to the third power of the object's distance from the center because for a sphere, the volume increases by the <u>cube</u> of its radius. But since the force felt acts from the center of the massive body, it ***decreases*** by the distance <u>squared</u>. When these functional variations are combined, they factor down and the force is seen to increase only as the <u>first power</u> of distance from the center, as the test mass climbs toward the surface. So Newton's formula is still being obeyed, both within and outside the massive body, even though the graph in figure 10 seems quite contrary.

One thing to remember in looking at these depictions of the Universe's apparent growth pattern is that the four-dimensional space-time construct is FLAT, not hyperbolic or elliptical. The curvatures noted in General Relativity relate to LOCAL effects produced by massive bodies. The Voidless Universe concepts proposed say that these local effects all integrate together to create a distributed gravity across the Cosmos that results in a flat space-time construct – which shows an **appearance** of a four-dimensional expansion curve.

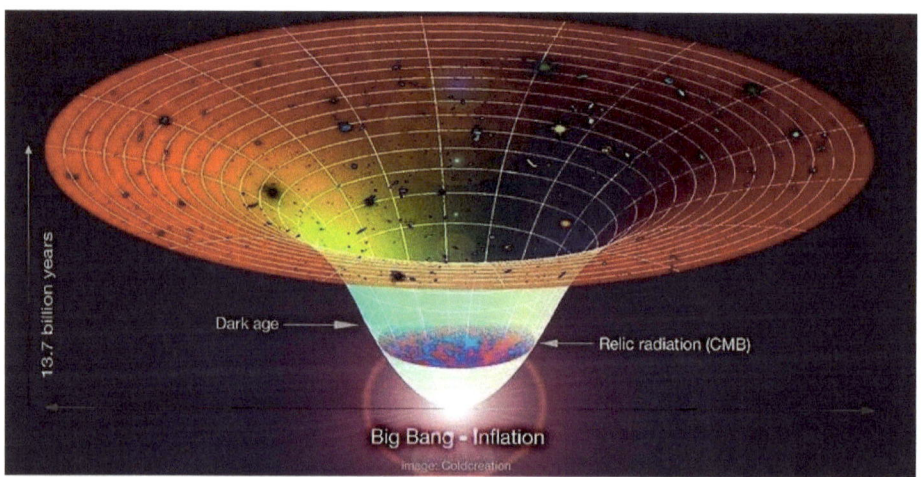

A number of rather esoteric studies of the Universe's curvature have been conducted and all indicate that a flat structure exists and that the Universe's density of matter and energy is equal to a value necessary to provide a flat space-time. This strongly implies that neither a future "Big Crunch" nor a cold, dark, empty Universe of perpetual expansion will occur. It also indicates that the fanciful ideas of higher-ordered dimensions [such as in String Theory] and curved space-time bending back on itself or vibrating "branes" twisting and intersecting are all just plain silly.

So what is really happening out there in the vast reaches of the Cosmos? We observe the Universe to be evenly distributed around us in all directions – with only the most minor variations. In a cosmologic sense, we sit in the center of a massive body (our Universe) that attracts us equally in all directions. At any distance we choose, the "shell" of matter being considered will exactly cancel out its effects on us.

Gravity – expressing the presence of mass in space, establishes a reference frame with respect to the world outside that mass. But the cancellation of these vectors is total. Thus, the presence of mass in the Universe establishes a reference frame of gravitational attraction to which its parts are excused from responding.

ASSERTION 3: Our gravitational reference frame defines a contraction that is being ignored. Distant celestial objects are not flying away from us in any organized manner. There is no actual expansion of space at all. The "Big Bang" is a mirage.

Any suggestion to overturn a theory as generally successful as Cosmologic Inflation (the "Big Bang") should, in addition to proposing alternative explanations, also cite at least one observation that the theory does not properly explain. In the case of the Big Bang, there are several:

1. Hubble Constant measurements providing 13.75 billion years for the Universe gives barely enough time to accommodate the age of the old reddish stars of elliptic galaxies and the globular clusters surrounding the Milky Way and other big spiral galaxies. While the evolution of galaxies and the globular clusters occurred simultaneously with the stars' coagulation from cold molecular clouds of dust and gas, another question still presents itself: Why don't the neighborhoods where most of these really old stars reside show any residual dust clouds left over from their stars' creation? The rare contradictory example of NGC 1316 is shown here, but the question still remains. Where are the clouds of dust and gas in all the other elliptical galaxies? How long did it take for them to be entirely consumed or dispersed?

NGC 1316

2. The Hubble Space Telescope's "Extreme Deep Field" snapshot of the Universe – reaching back twelve billion years into the past – shows galactic objects of rather mature shapes and colors. If the universe were really only two billion years old at that time, most of them should be small, irregular, bluish galaxies – with a smattering of recently-emerged spirals and ellipticals, perhaps. But the image shows a full range of both galactic types – many showing quite mature colors.

Hubble Extreme Deep Field

30

3. All elements whose nuclei are heavier than iron are predicted to be created only in the shells of supernova explosions, which then distribute those materials throughout their respective galaxies. [Those items of silver or gold jewelry you might wear on occasion are actually stardust – having come from the interior of a supernova that exploded before our sun was formed.] However, the observed frequency of supernovas is much too low to explain the amounts of iron and heavier elements detected by spectroscopic analysis of numerous celestial objects within the Milky Way and beyond.
4. The spectra of many ancient quasars (quasi-stellar sources) show them to possess "modern" levels of iron their atmospheres. If the rate of supernova explosions were much higher in the ancient past than now, argument 3 above could be resolved under a Big Bang scenario – but not the observation of these quasar iron levels. The presence of atmospheric iron indicates a second-generation celestial object. For ancient quasars to have "modern" levels of iron means that at least one generation of stars and galaxies has been missed out at a distance (and therefore back at a time) when nothing should exist – according to the "Big Bang".

ASSERTION 4: Cosmologic Inflation models simply do not allow enough time for well-documented processes, such as long-lived stars and galaxy formation, to have occurred in the Universe.

ALTERNATIVE EXPLANATIONS FOR THE TWO PRINCIPAL INFLATIONARY MODEL PHENOMENA

1. The "red shifting" of spectral absorption lines from their "laboratory" or reference wavelengths is what indicates to astronomers that objects are moving away from their telescopes. This technique is used to measure all manner of radial velocities in astronomy. Edwin Hubble's thoroughly remarkable discovery during the 1920s was that almost all galactic and similarly-sized objects in the Universe are moving away from the Milky Way in a very organized manner – at speeds which are in a virtual lock-step with their distance. This gave rise to the term "Cosmologic Red Shift" and decades of effort to accurately determine the value of the "Hubble Constant" began.

Though modern physicists have replaced Hubble's original "Doppler Effect" assumption with the "scale factor" expansion of General Relativity, the basic computation hasn't changed. The ratio of wavelengths between the original emitted light and the photons received is denoted by the dimensionless number "z".

$$\frac{\lambda_{Observed}}{\lambda_{Emitted}} = 1 + z \qquad [25]$$

It is computed thusly:

$$\frac{\lambda_{Observed}}{\lambda_{Emitted}} = 1 + z = \sqrt{\frac{1 + \frac{v}{c}}{1 - \frac{v}{c}}} \qquad [26]$$

where v is the source object's velocity and c is the speed of light.

For our sample of 12 galaxies, the average recessional speed was 1.3717 x 10^6 m/sec. The "z" value for this velocity is 0.004593. 1 + z gives us 1.0045859, which identifies the wavelength change due to the Hubble Constant expansion of the Universe for our average galaxy. The definition that relates photon energy to wavelength is:

$$E = \frac{\hbar c}{\lambda} \qquad [27]$$

where $\hbar$ is the reduced Planck's Constant, c is the speed of light, and λ is the wavelength of the subject photon.

We see that each photon arriving from our average galaxy (at a distance of 19.48 megaparsecs) has lost 4.3324 x 10^{-23} Joules of energy. This is an energy loss of 7.2048 x 10^{-47} Joules/m as the photon travels through space.

The Voidless Universe concept proposes that as photons traverse the reaches of space-time they lose energy to the generalized gravitational potential well of the Cosmos. This gravitational redshift is defined by General Relativity as:

$$\frac{\lambda}{\lambda_o} = 1 - G\frac{M}{c^2 r} \qquad [28]$$

This equation assumes that a photon will lose wavelength energy as it moves a distance r away from a mass M. For our "Big Bang" consideration we make adjustments: M will be the mass of the observable Universe and r is the distance the photon has traveled to reach us. M is estimated by multiplying the volume of the observable Universe by the density considered as "critical" to achieve the observed flat space-time [9.3 x 10^{-27} kg/m^3]. So the mass considered in "Big Bang" models is 3.348 x 10^{54} kg. This value considers "Dark Matter" and "Dark Energy" as two major components of its total (22% and 74% respectively).

Using this mass figure in equation [28] results in energy losses for photons coming from our twelve sample galaxies that average 18.96 times too high when compared to the Hubble's Law values. Substituting a Universal mass value of 5.2733% of that gives an exact six-digit conformance for our galaxy average – and a standard deviation of less than 0.2% among the twelve examples:

Galaxy Name	Distance [m]	Velocity [m/sec]	1+z (λ/λ_o)	Energy Δ Hubble [J]	Energy Δ Gravity [J]
NGC 4321	6.8815x10^{23}	1.570x10^6	1.005250	3.7823x10^{-23}	3.7844x10^{-23}
NGC 5505	1.8619x10^{24}	4.248x10^6	1.014270	1.3919x10^{-23}	1.3987x10^{-23}
NGC 5457	1.1703x10^{23}	2.670x10^5	1.000890	2.2294x10^{-22}	2.2253x10^{-22}
NGC 3623	3.3355x10^{23}	7.610x10^5	1.002540	7.8156x10^{-23}	7.8076x10^{-23}
NGC 3992	4.5847x10^{23}	1.046x10^6	1.003496	5.6834x10^{-23}	5.6803x10^{-23}
NGC 1365	7.2979x10^{23}	1.665x10^6	1.005569	3.5667x10^{-23}	3.5685x10^{-23}
NGC 300	6.3117x10^{22}	1.440x10^5	1.000480	4.1346x10^{-22}	4.1261x10^{-22}
NGC 5194	2.0381x10^{23}	4.650x10^5	1.001550	1.2797x10^{-22}	1.2778x10^{-22}
NGC 4548	2.1565x10^{23}	4.920x10^5	1.001644	1.2094x10^{-22}	1.2076x10^{-22}
NGC 1068	4.9660x10^{23}	1.113x10^6	1.003786	5.2462x10^{-23}	5.2441x10^{-23}
NGC 4254	1.0515x10^{24}	2.399x10^6	1.008036	2.4724x10^{-23}	2.7467x10^{-23}
NGC 4501	9.9496x10^{23}	2.270x10^6	1.007602	2.6135x10^{-23}	2.6174x10^{-23}
Average	6.0121x10^{23}	1.372x10^6	1.004593	4.3316x10^{-23}	4.3316x10^{-23}

Table 3

This demonstrates that Edwin Hubble's observations of galactic wavelength shifts were not caused by an expansion of space but rather by photons losing energy as they traveled through the Universe's gravitational potential well – caused strictly by "normal" baryonic matter.

ASSERTION 5: The "Cosmologic Red Shift", while truly cosmologic in origin, is **not** proof of a "Big Bang" (or any other inflationary) model of the Universe.

This also demonstrates that "Dark Matter" and "Dark Energy" are not necessary to explain the fictitious expansion of the Universe. They do not exist.

ASSERTION 6: "Dark Matter" and "Dark Energy" do not exist.

Another corollary here is that a photon moving through space will lose 7.2048 x 10^{-47} Joules of energy during each meter of its travel. From the edge of the observable Universe, this means a

33

loss of 2.3163 x 10⁻²⁰ Joules (59.21%) of its energy. Adding this effect to the inverse-square reduction of intensity noted in traditional Optics models easily explains "Olber's Paradox" regarding why the night sky is dark instead of white with starlight. No "Big Bang" or any other inflationary theory is necessary to explain it.

2. The other widely accepted "proof" of a Big Bang is the "2.7°K Background Radiation" first detected in 1965 as a mysterious "hiss" in radio telescopic observations. It is so pervasive and isotropic in strength that no variations in it were detected until data from the Cosmic Background Explorer (COBE) satellite were fully analyzed in 1992. The variations depicted below are estimated to be around 1x10⁻⁵ of the signal strength.

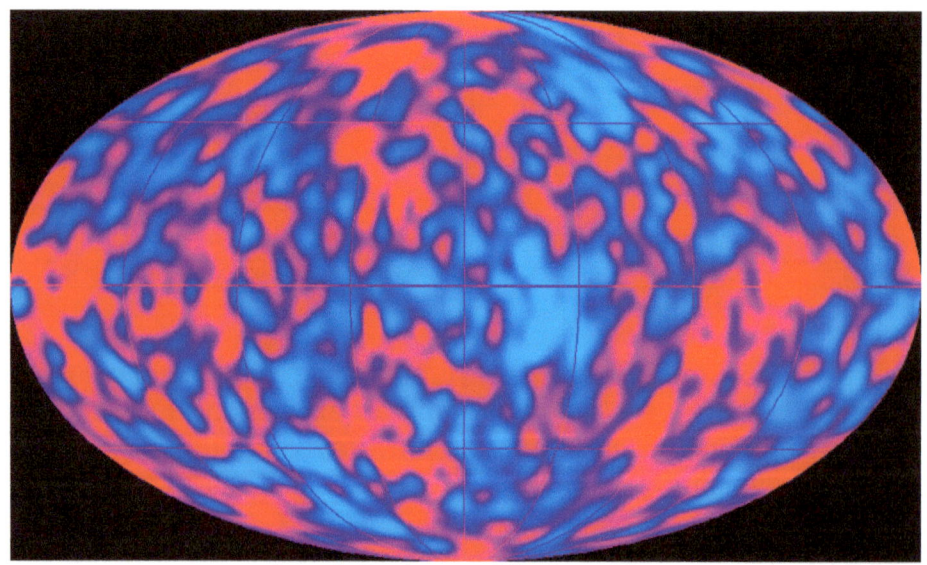

Cosmic Background Explorer Map

The effort to improve on this apparently very important image continued with the Wilkinson Microwave Anisotropy Probe (WMAP) launched in 2001. This satellite continues to measure the cosmic background radiation and adds increasing detail to the picture. On an approximately two-year cycle new, sharper images are provided to the community of astronomers and physicists. Here is the latest view:

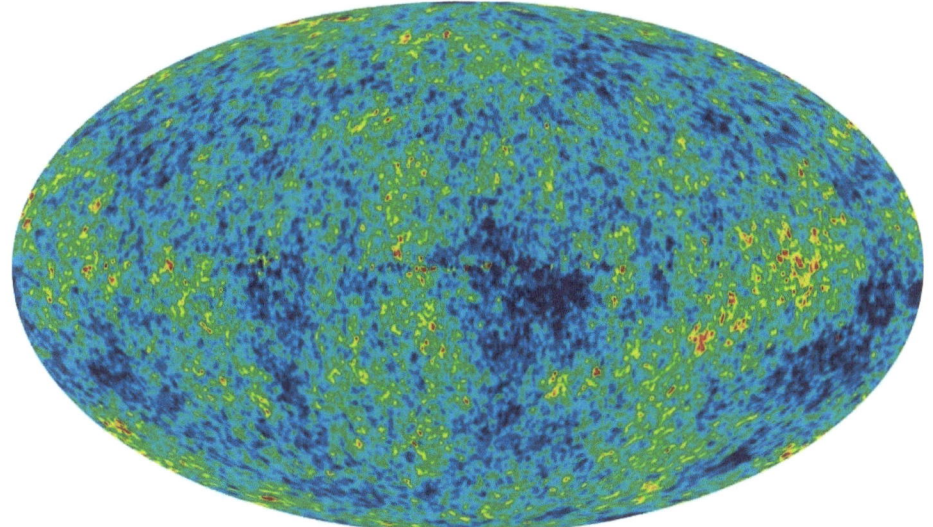

WMAP 2012 (9-year)

The WMAP features, which subtract out signals from the Milky Way, have from the start have looked very much like results from the Two Micron All-Sky Survey (2MASS) completed in 2003. The Voidless Universe concept proposes that this similarity occurs because the observed background radiation really stems from the random movement of matter through our observational reference frame – established by the Universe's gravity.

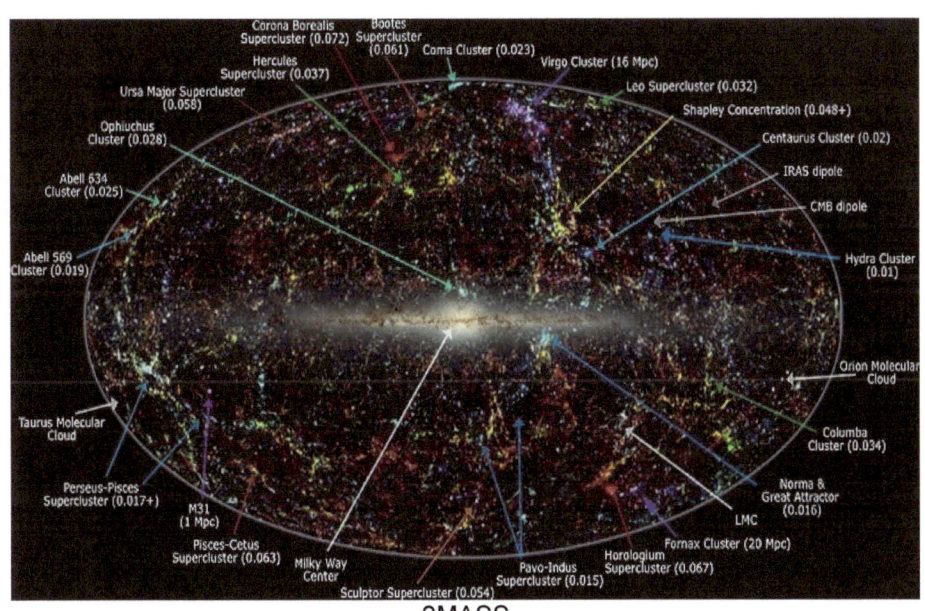

2MASS

The cosmic microwave background (CMB) radiation has been called the most perfect black body radiation observed in Nature. It seems to permeate all space at a thermal equivalent of 2.725° Centigrade above absolute zero, called degrees Kelvin. This means we can theoretically calculate the intrinsic intensity of the radiation from the energy density form of Planck's Law:

$$\frac{U}{V} = \frac{8\pi h v^3}{c^3} \frac{1}{e^{hv/kT} - 1}$$

[29]

$$= 1.6091 \text{ x } 10^{-25} \text{ Joules/}_{m^3}$$

One of the principal contentions of the Voidless Universe concept is that there is no organized cosmologic movement of the Universe's matter and energy – that both components of existence are free to move randomly, except where they may be subject to possible local constraints such as gravitational attraction to other bodies, magnetic fields, rotational motion around a parent body's center of gravity, etc.

To describe the true origin of the CMB radiation, we start with the baryonic density of the Universe: $5.2244 \text{ x } 10^{-28}$ kg/m^3. Entering this value into the formula for kinetic energy, we get:

$$\frac{E}{V} = (2.6122x10^{-28}) * v^2 \text{ Joules/}_{m^3}$$

[30]

Where v is the velocity of material moving randomly through space.

An initial exact match with the CMB energy density is achieved with a random material velocity of 25.6167 m/sec. But we see only that radiation streaming in our direction, so we multiply this energy density value by six. Our final random velocity of baryonic matter necessary to achieve the observed CMB energy density value then becomes 62.7477 m/sec.

Since the rotation speed of our average galaxy is almost 3600 times this value, it is obvious that more than enough random velocity of baryonic material is available in the Universe at large to provide the gravitational energy usually interpreted as the Cosmic Microwave Background radiation.

ASSERTION 7: The "2.7°K Background Radiation", while truly cosmologic in origin, is **not** proof of a "Big Bang" (or any other inflationary) model of the Universe.

THE EFFECT OF GALACTIC MERGERS

NGC 4038

In the above section on generalized galactic geometry, it was noted that galaxies evolve from irregular, through spiral and barred spiral, to elliptical forms. Since the Voidless Universe concept asserts that the Cosmos has existed for an indefinite, probably infinite, amount of time – the question of why spiral galaxies still exist surfaces once more. If galactic evolution progresses only toward the elliptical form, all galaxies should have become elliptical by now. As noted earlier, around 35% of the Universe's visible matter is collected in spiral or irregular galaxies, so there must be some cosmic process that disrupts or reverses this evolution – driving galaxies back towards the more "primitive" galactic shapes.

The impressive scene just above (NGC 4038) captures two spiral galaxies in the midst of joining into one elliptical galaxy. Immense bluish fields of star creation are visible on the left and vast clouds of smaller, reddish stars hover off to the right of the main action in the center. This majestic chaos isn't as violent as it might seem to be at first glance. Statistics show that very few, if any, of the stars involved in these merging systems will ever collide with another. Even the super-massive black holes suspected to reside at the centers of the original galaxies will not be likely to meet for many millions of years. What does collide are the vast clouds of dust and gas the galaxies contain. Billions of massive stars are created, burn furiously for millions of years, and then explode – violently casting heavier elements into the space around them. Many more billions of smaller stars are also created within the compressing clouds. These don't explode, but quietly shine for billions of years while the galactic chaos around them subsides into a stellar-level form of viscosity and they nestle into the very dense "stellar city" of an elliptical galaxy.

But all is not really quiet there. Stars in elliptical galaxies exhibit beehive-like stellar motions. While the overall galactic structure seems sedate, the individual stars swarm furiously within it.

At the galactic level, this merger may require a billion years or more to settle down into an elliptical shape similar to that of NGC 1316, shown on page 30. Several billion more years may be required to consume or disperse the vast clouds of residual gas left over from the original tumult of the merger. While that expanse of time presents a problem for the "Big Bang" or any inflationary model of Cosmology, the issue for the Voidless Universe concept to answer is: Why haven't all these mergers been completed long ago? Why isn't all visible matter already collected into "stable-looking" elliptical galaxies?

The answer to the mystery is that elliptical galaxies themselves are also subject to the same merging process that created them. When two or more of these immense systems migrating through space pass close enough to each other to start exerting significant gravitational attraction, it is rarely a head-on collision. The same picturesque process of galactic disassembly occurs and their material is strewn through space as depicted in the example of NGC 4038. During these mergers, however, the much higher stellar densities of elliptical galaxies do cause their stars to physically disrupt the structural integrity of other stars passing close by. The "distributed source gravity" argument of the Voidless Universe concept describes how the swarming stars of the colliding elliptical galaxies will shred their stellar neighbors while they all fling each other's material across the Cosmos. The galaxies' angular momenta around a mutual center of gravity adds to the chaotic velocities of the individual stars to ensure that a noticeable proportion of them escape the local neighborhood altogether.

The evolution of galaxies is seemingly thrown backward in the direction of its original components. The ratio of 65% elliptical to 35% spiral and irregular galaxies is the "steady state" result of the mutual evolutionary processes of galactic formation. In generalizing the Universe's galactic geometry above, the Voidless Universe concept turned Hubble's "Tuning Fork" diagram into the single strip of Figure 7. We now see that the strip turns around on itself into a belt of galactic evolution.

Assertion 8: The evolution of galactic structure is not singular in direction. Spiral galaxies (barred or not) evolve into ellipticals, but elliptical galaxies themselves continue on, to eventually re-evolve into their various components.

LIGHT'S SPEED AS AN ACOUSTIC VELOCITY

If we consider the Universe to be a continuous medium of very low density – rather than empty space containing dense knots of matter (mathematically treated as point sources), we can explain observations of high-speed jets of material being ejected from the super-massive black holes at the center of numerous galaxies. As noted at the beginning of this work, the task of Astronomy is to explain observations through development of consistent theories, not to contort observations to fit false explanations. The simple solution here is that a value usually interpreted as a "fundamental constant" by physicists is really a *function* of the medium.

When we consider the Universe as a continuous medium, light's speed can be redefined as celerity, or acoustic velocity. A bit of ironic coincidence appears here in that the commonly used symbols for the speed of light and for celerity are the same – a lower-case letter "c". The formula for determining numeric values for celerity is:

$$c = \sqrt{\frac{B}{\delta}}$$ [31]

where δ = density of the medium and

$$B = \frac{\Delta P}{-\Delta V / V}$$ [32]

The "B" in this formula represents a quantity known as the "Bulk Modulus". It is a property of the medium under consideration, in this case, the Universe. As *its* formula shows, it is defined as the incremental shrinkage of volume resulting from an increase in pressure.

Since the speed of light has been accurately measured and we have an estimate of the "critical density" of matter in the Universe, calculating a numeric value for this "Bulk Modulus of the Cosmos" should be child's play. But let's wait a moment. Einstein said that mass and energy have an equivalent relationship:

$$E = mc^2 \qquad [33]$$

From the acoustic velocity formula:

$$B = \delta * c^2 \qquad [34]$$

Since density is defined as the mass of something divided by the volume in which it resides, Einstein's definition of the energy-mass relationship shows us that, in a cosmologic sense, the "Bulk Modulus of Space" will be equivalent to the mass-energy density of the Universe.

$$B = \frac{m}{V} c^2 \text{ or } B = \frac{E}{V} \qquad [35]$$

So what does this mean? It means that the "compressibility" of space, or the scalar curvature of the Universe's space-time continuum, is definable as the density of the mass and energy it contains. It means the speed of light can be numerically determined as a function of that density. It is *not* a "universal constant", mystically determined by whatever established the Cosmos. If the Universe were denser, light's speed would be lower. If the Universe were less dense, the speed would be higher. Taken to an extreme, it means that, in either "Big Bang" or black hole scenarios, the speed of light tends toward zero. This in turn would mean that exchanges of mass for energy (and back again) could take place with a high degree of ease. The current feeling among astronomers and physicists that the "normal laws of physics don't apply" or can't be described inside a black hole is an error. What is different in such high material density environments is that matter and energy exchange their states more freely – period.

Indeed, at the subatomic level, the equivalence of matter and energy is freely discussed as a topic of instruction. The masses of those particles are often measured in kilograms and then equally as often expressed in units of electron volts. This is fully appropriate because the density of nuclear material (at 2.3 x 10^{17} kg/m^3) is so high that the speed of light within the nuclear environment slows to a value low enough to facilitate the easy conversion of matter to energy – and back again.

The current question is: Is the Universe's gravity strong enough to "squeeze" its own volume to a point where the "acoustic velocity of space" is equal to light's speed? We can demonstrate this by showing that the bulk modulus, integrated across the outer boundary of the Universe, is numerically equal to the gravity web arising from the mass within it. Let's start with Gauss' Divergence Theorem, which describes the flux of a force through a surface area:

$$\text{Flux} = \int F \bullet n \bullet dS = \oiint div F dV \qquad [36]$$

where, for gravity,

$$F = \int -\frac{GMr}{r^3} \qquad [37]$$

and n is the normal vector perpendicular to surface S.

$$\text{Flux} = \oiint -\frac{GMr}{\left|r^3\right|} \bullet -\frac{r}{\left|r\right|} dS \qquad [38]$$

$$= GM \iint \frac{r \bullet r}{r^4} dS \qquad [39]$$

$$= GM \iint \frac{1}{r^2} dS \qquad [40]$$

Over a sphere (We observe the Universe in a spherical sense because we will always have only one location for viewing it.), dS evaluates as **4πr²**. So:

$$Flux = 4\pi GM \qquad [41]$$

The variable in this formula is the mass, **M** – determined by the Cosmos' density times its volume. Though the flux might still appear to have a radial distance component ($M = \frac{4}{3} * \pi * \delta * r^3$), now its value is simply measured by taking **r** as the radius of the observable Universe. Evaluating:

$$\text{Flux} = 4 * \pi * (6.6738 \times 10^{-11}) * (1.6546 \times 10^{53}) \qquad [42]$$
$$= 1.3876 \times 10^{44} \text{ nt or, per unit area:}$$
$$= 5.9207 \times 10^{-11} \text{ }^{nt}/_{m^2}.$$

The unit flux developed from the celerity formula [31] where $\delta = 4.9042 \times 10^{-28}$ $^{kg}/_{m^3}$ gives a value of 4.4076×10^{-11} $^{nt}/_{m^2}.$

In this first rough calculation of the Bulk Modulus we get an energy flux that is 34.3% smaller than the flux of gravitational force through the surface of the Cosmos. What will it take to get an exact agreement? The speed of light has been very accurately measured. The mass initially used in this computation is derived from the density required to support the observed "flatness" of space-time. The real variable here is the radius of the observable Universe.

We get full agreement [to within six significant digits] between the "celerity" computation and our calculation of gravity's flux through the surface of the Cosmos with a radius of 3.215×10^{26} meters for the observable Universe. The Cosmologic Bulk Modulus then, is $B = 4.4076 \times 10^{-11}$ $^{nt}/_{m^2}.$

What this means is that gravity is strong enough to establish an acoustic velocity equal to the speed of light as long as its radius is equal to or greater than 33.982 billion lightyears. Cosmologic views of "Big Bang" and other inflationary models currently estimate a radius of the observable universe as 46.5 billion lightyears. This demonstrates that the value of the speed of light is actually established by the acoustic velocity of the Universe.

ASSERTION 9: What has been called the "speed of light in a vacuum" is actually the acoustic velocity (the celerity) of the Universe. While it is the highest speed usually attained by natural processes, it is not the highest speed possible.

Plasma jets streaming from super-massive black holes in the center of large galaxies or quasars are often ejected at speeds much beyond light speed. This discussion has established a straightforward explanation for these observations without resorting to any tricks or the tortured logic of "special case" solutions.

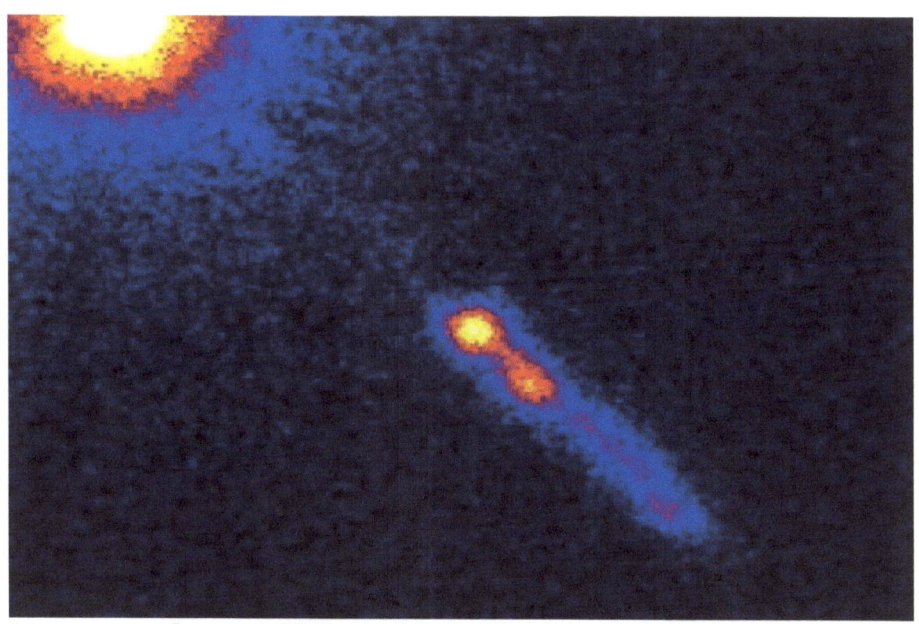

Quasar 3C273 and its superluminal jet (~ 9.6 x c)

A PARADOX EXPLAINED

Another "problem" this approach solves for Physics is the dichotomy arising from trying to fully explain the nature of light itself. [Is it a "wave" or is it a "particle"?] Consider a single photon streaming across the Cosmos. What is it going to do when it interacts with one of our many favorite experiments? Wave-oriented experiments consistently display wave-like results, while particle-oriented setups always provide photon-type observations.

This is known as the "Wave-Particle Duality Paradox". The concept states that ALL particles (and photons, too) exhibit both wave and particle properties. It is an observation that has been difficult for theoretical physicists to explain for nearly a century (at this writing). The explanation, actually, is simple. The Voidless Universe creates a seamless web of interaction to which all of its components – of whatever size, energy, or momentum – respond. As entities of any type move (and they ALL do move, relative to the whole) they are influenced by the whole. Whether defined by probability or called by Fate, the pieces of the Universe are tugged on by the world behind them. This "wake-like" effect shapes the course of their progress through the Cosmos. But as each surges forward, whether through the low-density medium of outer space or the thickness of a glassy lens or magnetic field, it is affected by the jelly-like consistency of the whole Cosmos through which it moves.

Gravity establishes energy potential fields throughout space – without interruption. Whether or not photons have mass they do possess measurable momentum. It is a photon's momentum as it travels towards us which is subjected to the acceleration field of the Universe.

ASSERTION 10: It is the universal medium's wake-like effect on them that creates the wave behavior observed in experiments performed on the quantum communities of photons and sub-atomic particles. It is the entanglement of these wakes that transfers energy from particle to particle, not suddenly-appearing leptons.

DISTRIBUTED GRAVITY, GENERALLY SPEAKING

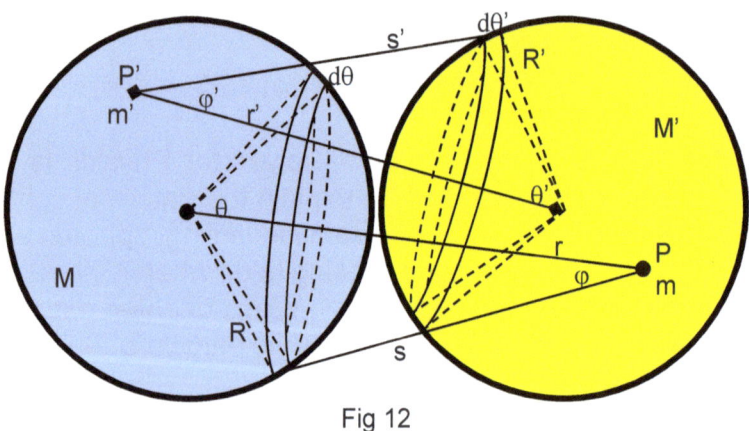

Fig 12

The arguments presented in relation to astronomical situations have shown that gravity cannot always be assumed to act strictly as a "point-to-point" source. Indeed, Newton's original derivation shows that it **never** acts in that manner. In the following studies, the angle φ and distance s from the original integral formula will have values dependent upon the circumstances being reviewed.

$$F = -Gm\int \frac{\cos\varphi * dM}{s^2}$$ [43]

Figure 12 just above uses his integral formula in a case where two masses of the same size interact at close range. The points P and P' in each sphere indicate the reference spots where m and m' were located in the original derivation. [Look back to Figure 1] We start by substituting values provided by the spherical geometry involved:

$$F = -\frac{GMm}{rR} \int \frac{\cos\varphi * ds}{s}$$

[44]

For two equal-size spheres as in the figure, the angle φ will range from 90° down to a lower limit of 18.43° and the distance s will range from a maximum of 4R down to 0.

$$F = -\frac{GMm}{R^2} * \sin\varphi \Big|_{18.43}^{90} * \ln s \Big|_0^{4R}$$

[45]

$$F = -\frac{GMm}{R^2} * 0.341886 * \ln s \Big|_0^{4R}$$

[46]

The "usual" computation for gravity is shown in the first term above. The second term is a multiplier – evaluated for this spherical case – that must be considered in all cases where the sources of gravity are not sufficiently separated to behave as point sources. [Other geometric examples were drawn in our earlier discussions of galactic gravity.] Students of Mathematics know that a value's natural logarithm approaches negative infinity as the value itself goes toward zero. For Physics, this means that within two massive bodies so aligned the force of gravity grows to nearly infinite strength in those portions that come close to actually touching. In Astronomy, the difference in gravity's strength from one side of an orbiting body to the other can disrupt the structural integrity of planets and moons coming too close to their parent bodies – and comets approaching too close to a planet as they pass by, as detailed in theories about the Roche Limit. [Examples: The rings of Saturn and the disruption of Comet Shoemaker-Levy 9 on the orbit before it impacted Jupiter.] Continuing our study of gravity to ever-smaller ranges in the Voidless Universe, we'll stay with spheres but go back a step to the formula:

$$F = -\frac{GMm\sin\varphi}{rR} \ln s$$

[47]

To study gravity's effects at such close ranges, we want to look at potential energy. The equation we're looking for is derived thusly:

$$U = -\int Fds = -\frac{GMm\sin\varphi}{R}\int \ln s\,ds \qquad [48]$$

$$= -\frac{GMm}{R} * \left[\left(\sin\varphi * s\ln s - s\right)\right]^{s} \qquad [49]$$

Figure 12 shows that we have two masses to be considered, so our final gravitational energy formula becomes:

$$= -G\left[\left[\frac{Mm}{R} * \left(\sin\varphi * s\ln s - s\right)\right] + \left[\frac{M'm'}{R'} * \left(\sin\varphi' * s'\ln s' - s'\right)\right]\right] \qquad [50]$$

Parametric review of the numeric results of formula [50] reveals that as the differential distance **s** gets smaller and smaller, the **(s * ln s – s)** term diminishes toward zero even more rapidly.

So the distributed-mass computation of gravity is maximized at the outside edges of the bodies but minimized (to zero) at any point approaching actual contact. This behavior provides a rational explanation for gravity's disappearance in the nuclear realm – where the simplistic interpretation of Newton's famous formula says it should be reaching its maximum strength and influence. Physicists are correct in assuming that gravity doesn't operate in the nuclear realm – but rather than just dismissing it, these formulas provide a convincing explanation.

ONE MORE THING ABOUT GRAVITY

Most of the proposals in this work drive to quantify each topic as much as possible. Now at the end, a qualitative issue arises: <u>Which physical entity is it that gravity is actually attracting?</u>

We've discussed how gravity has always been measured as a force attracting masses towards each other. While pondering several effects his discovery might have in the natural world, Newton and his cohorts in the late 17th century erroneously assumed light rays had mass in attempts to describe wavelength changes that might be attributed to gravity's influence.

Later on when light was declared to be massless, gravity's theoretical effect on it became a true mystery. By the time those influences were finally measured accurately, a new vision of gravity (Einstein's General Relativity) had described the geometric environment that mass creates in defining itself to the Cosmos and we started measuring gravity in terms of potential energy wells. [Einstein's computation of wavelength shifting due to gravity turned out to be exactly twice the value of Newton's estimate.]

Light is correctly brought into discussions of gravity by noting that while photons have no mass they do possess momentum and associated values of energy. In an earlier work (Special Relativity) Einstein defined the equivalence between mass and energy:

$$E = mc^2 \qquad [51]$$

In another area of Modern Physics, DeBrolie's definition of wavelength equivalents for massive bodies (such as protons, neutrons, electrons and the like) and the "Wave-Particle Duality Paradox" principle take us a step further towards a final conclusion about gravity.

A primary principle of the Voidless Universe concept is that the energy-matter mixture of the Universe spreads its gravitational influence all through the Cosmos, without limit – in extent or scale. It is further postulated that Nature's rules are simple even though they may be exercised in complex ways, depending on circumstances. This brings us to conclude that gravity operates on energy – not mass.

The Earth's gravitational field attracts the <u>energy</u> equivalent of your body's mass towards the center of its potential energy well.

In this light, a new equivalent formula for Newton's version of gravitational force might be rewritten as:

$$F = -G\frac{Mm}{r^2} = -G\frac{Ee}{c^4 r^2} \qquad [52]$$

And its corresponding formula for gravitational potential energy will become:

$$U = -G\frac{Ee}{c^4 r} \qquad [53]$$

In these formulas the numeric value for the G term is: 3.7438×10^{19} MeV-m/kg^2.

So as our final statement on gravity:

ASSERTION 11: Gravity operates as energy attracting energy, not mass attracting mass.

WHAT THE "VOIDLESS UNIVERSE" MEANS TO SUB-ATOMICS AND THE "GRAND THEORY OF EVERYTHING"

An effort being expended at <u>great</u> expense across the scientific world is the pervasive reflex among many physicists to deconstruct the components of the Universe at its smallest level, hoping to drive toward an explanation of the Universe's four principal forces as expressions of one unified force. The thinking is that at some extremely high level of energy (assumed to be present at the beginning of the "Big Bang" period) all four forces must have had one unified expression – which differentiated itself at the lower energies that occurred when the Universe expanded.

These forces are:

1. Gravity.
2. The Strong Interaction, which creates protons and neutrons – while its residual effects create atomic nuclei.
3. The Weak Interaction, which creates nuclear radioactivity and influences various other sub-atomic changes.
4. Electromagnetism.

The "Grand Theory of Everything" proposes that as the newborn universe appeared it very rapidly evolved, by expansion and cooling, through phases where first gravity, then the strong force, then the weak force, and finally electricity and magnetism all separated from each other into their current expressions.

Current scientific equipment can produce energies sufficient to emulate a merger of Electromagnetism and the Weak Interaction to mimic an "Electro-Weak" force. Theory indicates that energies necessary for the other steps are far beyond anything mankind will ever be able to achieve, so they will forever remain improvable speculations.

One of the tenets of Modern Physics is that forces are transmitted between quantum particles through the exchange of "force-carrier" particles. The "gauge bosons" of the "Standard Model of Particle Physics" are defined as entities that transfer energy between "Hadrons" and "Fermions". So far this theoretical construct has defined a "zoo" of 61 species of particles – and promises more. The reason this semblance of order appears is as follows:

When a fragile stemware glass is thrown on the floor, the possible configurations of its pieces, when finally at rest, are nearly endless. But as the glass is smashed down harder and harder, the pieces get smaller and smaller – more closely approaching the sand from which the glass originally came. The logic of quantum physics dictates that as the original items being destroyed are smaller and smaller, the resulting collections of fragments will start to achieve certain patterns of definition. So we have the "Standard Model" taking shape.

In the atom-smashing process that is used as its primary tool, the resulting products of each collision really exist across a continuum of energy. Physicists then go through a rather laborious process of "normalization" to determine into which bucket of the "Standard Model" a particular observation might fit when the particle is imagined to be actually "at rest". When they find one that doesn't fit into any of their previously-defined boxes, they hail it as a new discovery. In the future, with wider spectrums of energy provided by ever more powerful equipment that successive generations of physicists can dream up, the game could be played endlessly – with little meaningful result.

Under the "Voidless Universe" concept, the energy transfers between the original entities are interpreted as web-like interactions between the "wakes" of the subjects' mass-energy bundles as they move toward and past each other in the Cosmos. Defining a menagerie of transfer particles is not necessary.

One of the principal concerns of the "Voidless Universe" concept is that astronomers and physicists have ignored Newton's "warning label" about using his final formula for gravity. We've discussed the problems this has caused for Astronomy – providing solutions for those situations. As noted just above, the disappearance of gravity in the nuclear realm is a natural outgrowth of the proper application of Newton's integral computation process. The further understanding that gravity operates on energy, not mass, shows that the efforts to squeeze gravity into a "quantum box" are unnecessary. [Searches for the "graviton" should be terminated.] This indicates that the highly-desired bridge between gravity theory (usually expressed these days as General Relativity) and Quantum Mechanics is also unnecessary.

The "Grand Theory of Everything" belief in "One Force" is very attractive –but it is not really necessary for the effective functioning of the Cosmos. The "Voidless" concept refutes that the "Big Bang" inflation of the Universe ever occurred. So we must conclude that the energies necessary to realize the theoretical unification of all forces may never have existed in Nature. The "Grand Theory of Everything" that the "Voidless Universe" contemplates is that the Cosmos exists as it exists – for its own sake, not necessarily for our convenience of understanding.

CONCLUSION

At the outset of this work it was noted that Astronomy is an <u>observational</u> science, not an experimental one. Astrophysical theories should be developed to explain observations – as a first-order effort. Theories should not be compounded upon each other simply to "prove" any of them correct – particularly when subsequent observations point out errors or inconsistencies.

My personal belief that the "Big Bang", Cosmic Inflation, or whatever it might be called, is incorrect – is a view I have held for more than fifty years so far. I review the considerations presented here whenever new relevant astronomical information is publicly provided. I have yet to find any observation that refutes the conclusions drawn in this work.

The general problems for Astronomy discussed that relate to galactic gravity and cosmology are:

1. Spiral galaxies, such as the one we're in, shouldn't exist.
2. The arms of spiral galaxies move at speed distributions that have been inexplicable to usual methods of computation.
3. "Dark matter" was first proposed as means to explain these first two issues – but it has never been observed.
4. Inflationary models of the Universe do not provide enough time for many well-documented astronomical processes to occur.

Two additional topics raised were:

1. The speed of light is determined by conditions of the Universe and is not arbitrarily set through any cosmic mystery.
2. Gravity is, in truth, a force where energy attracts energy, not mass attracting mass. The equivalence between mass and energy pointed out in Special Relativity reveals why this effect has been effectively masked to our usual understanding.

It remains to be seen whether or not these additional considerations bring us closer to a "Theory of Everything". Generally speaking, the "Voidless Universe" doesn't need one.

On a more specific basis, the new notions proposed in this description of the Voidless Universe are numerous. Let's list them:

ASSERTION 1: The material in galactic spiral arms is more significantly influenced by the matter near it in the spiral disk than by the galactic core. A galaxy's spiral arms (and its associated bars) are actually a separate gravitational system.

ASSERTION 2: Huge quantities of mysterious "dark matter" are not required for spiral galaxies to retain their material. Their massive cores, extensive luminous spiral arms and thin envelopes of normal non-luminous matter are sufficient to constrain all their material.

ASSERTION 3: Our gravitational reference frame defines a contraction that is being ignored. Distant celestial objects are not flying away from us in any organized manner. There is no actual expansion of space at all. The "Big Bang" is a mirage.

ASSERTION 4: Cosmologic Inflation models simply do not allow enough time for well-documented processes, such as long-lived stars and galaxy formation, to have occurred in the Universe.

ASSERTION 5: The "Cosmologic Red Shift", while truly cosmologic in origin, is *not* proof of a "Big Bang" (or any other inflationary) model of the Universe.

ASSERTION 6: "Dark Mass" and "Dark Energy" do not exist.

ASSERTION 7: The "2.7°K Background Radiation", while truly cosmologic in origin, is *not* proof of a "Big Bang" (or any other inflationary) model of the Universe.

Assertion 8: The evolution of galactic structure is not singular in direction. Spiral galaxies (barred or not) evolve into ellipticals, but elliptical galaxies themselves continue on, to eventually re-evolve into their various components.

ASSERTION 9: What has been called the "speed of light in a vacuum" is actually the acoustic velocity (the celerity) of the Universe. While it is the highest speed usually attained by natural processes, it is not the highest speed possible.

ASSERTION 10: It is the universal medium's wake-like effect on them that creates the wave behavior observed in experiments performed on the quantum communities of photons and sub-atomic particles. It is the entanglement of these wakes that transfers energy from particle to particle, not suddenly-appearing leptons.

<mark>ASSERTION 11</mark>: Gravity operates as energy attracting energy, not mass attracting mass.

The eleven assertions summarized just above and the new concepts proposed in this work show that many on-going experiments in Astronomy and Physics to find "missing" or "dark" matter / energy / whatever are misguided and might well be terminated.

Another major point made is that Edwin Hubble's initial "Doppler Effect" interpretation of the galactic redshifts he measured was wrong. The gravitational redshift predicted by General Relativity was known at the time. Gravity's effect on bending light rays had been observed a few years earlier. Hubble should have investigated what effects Relativity might have played in his observations. Almost all astronomers and physicists involved in the field ever since have compounded the error by assuming that these effects are actually occurring in a grotesquely expanding Universe.

Some philosophical discussions in science have lately noted the value of "Occam's Razor" as a guiding principle – although William of Ockham himself probably never phrased it as concisely as recently interpreted by many people. Wikipedia defines it as: "…among competing hypotheses, the one that makes the fewest assumptions should be selected."

The worldview proposed by the "Big Bang" and the mathematical gymnastics brought to bear in support of it are utterly confounding in their complexity – which should warn many of the individuals involved that something is very wrong with the basic idea. Saying that the entire Universe is expanding at least three times faster than the speed of light because that will justify the numbers observed is a preposterous proposition.

The term "Cosmology" is defined as the "general science or philosophy of the Universe". In that light, every particle or location in Existence will have some corresponding viewpoint of "the Whole" – and will also respond to influences from that Entirety.

The discussions above reveal that there is **no overall organized** scenario to which "the Whole" is responding. The components of the Universe exist as they are and move the way they do for their own sake, not ours. Each alternative presented here to a

"Big Bang" counterpart effectively asserts that it is a more physically-normal phenomenon.

The revolutionary viewpoints proposed above go further than just trying to refute an implausible, though popular, theory. At its most basic, the Voidless Universe concept stresses that Nature fills a continuous four-dimensional reference frame and its space-time structure cannot be considered separately from the mass and energy that it contains. The Cosmos is not constrained by any of its dimensions. Creation had a beginning that is beyond definition or measurement and extends without limit in either space or time. The Universe *is as it is* at every cosmic point – unchanging and constant in the Whole – and Voidless.